Every Number Is Special

Boyd Henry

DALE
SEYMOUR
PUBLICATIONS

PREFACE AND OVERTURE

The author wishes to express his profound appreciation to the team of Gilbert and Sullivan for providing the logic needed to "prove" the theorem EVERY NUMER IS SPECIAL.

Perhaps this acknowledgment needs an explanation.

In their operetta RUDDIGORE, OR THE WITCH'S CURSE, through a complicated turn of events, Sir Ruthven Murgatroyd is appalled to find that he has inherited an infelicitous family curse. This malediction requires him to commit a crime each and every day of his life or suffer death by torture on the very day that his crimes cease. Each previous Lord of Ruddigore has so died. As it happens, Sir Ruthven is basically a thoroughly good chap. Because the curse, most foul, condemns him to be bad, he finds himself facing an unbearable dilemma. But, as any devotee of Gilbert and Sullivan instinctively knows, neither Sir Ruthven, nor the audience that shares with him his deep anguish, is condemned to despair forever... or even beyond the end of the performance. Surely, Sir Ruthven can be expected to hit upon a most happy solution. Indeed, he does. For, if he fails to commit his daily crime, that lack of action is tantamount to suicide. But, AHA! Suicide is in itself a crime! Thus he can remain thoroughly good and at the same time commit his required daily crime. What a marvelous solution! How very moral!!

Now to the theorem. EVERY NUMBER IS SPECIAL!

It is with Sir Ruthven's incontrovertible logic that we prove this theorem. (We need to issue a disclaimer at this point. This theorem refers only to positive whole numbers. It was never intended to include negatives, irrationals, imaginaries, transcendentals, or other nonconforming members of the subculture of numbers with a deviant life style.)

Proof: Assume for the moment that there do exist some numbers that aren't special. Some may even be thoroughly nonspecial. These numbers are numbers of a nonempty set of nonspecial numbers. Because they are positive whole numbers, the well-ordering principle requires that the set contain a least element, say *N*. Thus *N* lives with the curse of being the least non-special number. As we bring this "proof" to a close, we hit upon a most happy solution. For, the very fact that *N* is the least nonspecial number is tantamount to saying that it is special indeed. So, AHA! *N* is not nonspecial after all. We have only to repeat this argument as often as necessary to remove, one-by-one, any other numbers thought to be cursed with nonspecialness. Hence, EVERY NUMBER IS SPECIAL. Q.E.D.

Readers should find this book to be a source of mathematical amazement, amusement, and challenge. After reading a fact about an interesting number, readers can select some numbers of their own choosing and check to determine if the claim seems to hold. Then they should try to prove the fact. Many facts can be easily proven with simple algebra. Others require a computer search. For example, probably the only way we can prove that 26 is the only number "squeezed" between a square (25) and a cube (27)--at least the only number less than 10 quadrillion--is to ask the computer to look around for any other such numbers.

Finally, should you find something interesting about other numbers, I would greatly appreciate it if you would please take the time and trouble to send those discoveries to me so I can add them to my collection.

Boyd Henry
The College of Idaho
2112 Cleveland Boulevard
Caldwell, Idaho 83605

INTRODUCTION

This book is not for everyone.

Not everyone enjoys good music. Not everyone thrills to an exciting football game. Not everyone loves great literature. Not everyone is intrigued by history. Not everyone delights in discovering surprises about ordinary counting numbers. But some do enjoy good music; some are thrilled by an exciting football game; some do love to read; some are intrigued by our past; and some do find great delight in discovering number surprises.

This book is for the last group.

Teachers can use this book as enrichment material in a variety of ways, depending upon their students' mathematical maturity and curiosity.

Different interesting facts require different handling. In some cases, the student may enjoy verifying the fact with a variety of examples. In other cases, he or she may wish to prove the allegation. In still others, a computer search may be necessary to confirm the stated fact. For example, it might be extremely interesting to ponder the fact that 26 is the only number (less than 10 quadrillion, at least) to be "squeezed" between a square, 25, and a cube, 27. That's a very nice property. But there is little that one can do to verify the statement without a computer. With a computer, however, one might extend the search beyond 10 quadrillion. (One might even check to make certain that our computer didn't overlook other such numbers because of a carelessly written program.) On the other hand, a simple hand calculator is all that's necessary to pursue the interesting fact about 17. (The difference between any square number and the nearest multiple of 17 is always a power of 2 [or 0].) With a calculator students can verify this fact by selecting a large variety of squares to test the statement. Some students may be interested in trying to prove the statement. (Hint: Investigate the pattern of square numbers, mod 17.)

Some teachers may wish to feature an "interesting number of the week." Each week one of the interesting number facts can be posted on a bulletin board (or otherwise publicized) for students to ponder, to verify, or to prove.

It is our ultimate hope, however, that capable students will be inspired to find for themselves interesting facts about numbers. We list herein something interesting about every number from 0 to 150. We had to stop somewhere! Can you or your students find anything interesting about 151 or 152? Can your class extend the list to 200? To 300?

There are two ways in which one might proceed in finding new number facts. Usually, the easier of the two is to pick the fact first and then find the number that fits the fact. For example, one might wish to identify the smallest number that is divisible by every whole number from 1 through 30. (That number is 2,329,089,562,800.) The second, generally more challenging activity is to pick a number--say 179--and try to find something interesting about it.

To summarize, here are some of the ways to use this book.

1. Sit back, read, and enjoy. Believe everything in the book without question. (Would we lie to you?)

2. Verify a fact with a hand calculator using many different choices of appropriate numbers.

3. Prove the fact algebraically or otherwise.

4. Perform a computer search for additional numbers with the same property or verify the results of our computer search.

5. Make up a fact and find a number that fits that fact.

6. Pick any whole number and discover an interesting fact about it.

Of course, we welcome any correspondence from our readers with respect to questions or new facts that they have discovered.

B. H.

0

0 is a troublemaker! It requires special handling that no other number requires.

0 is the only number that's neither positive nor negative.

0 is the only number that can never be used as a divisor in a division.

0 is the only number that must be one of the factors if it is the product of two numbers.

0 is the only number that doesn't affect the sum when used as an addend or affect the difference when used as a subtrahend.

0 is the only number that doesn't have a symbol in most ancient notation systems. There was no numeral for 0 in the Roman, Egyptian, Greek, Babylonian, and Chinese numeration systems.

0 is the only number we use to "round off" other numbers. We round 93,123,456 to 93,000,000, but we would never think of rounding it to 93,111,111.

0 is the only number that is frequently incorrectly called a letter of the English alphabet. Its name is zero, not oh.

1

1 is the only counting number that is neither prime nor composite.

2

2 is the only even prime.

2 is the only solution to the equations

$$n + n = n \times n = n^n$$

3

Every even power of any integer is either a multiple of 3 or exceeds a multiple of 3 by 1.

Examples: $2^6 = 64 = (3 \times 21) + 1$

$14^4 = 38{,}416 = (3 \times 12{,}805) + 1$

No even power of an integer is ever 1 less than a multiple of 3.

4

If any whole number greater than 1 is raised to the 4th power and the result is increased by 4, the answer will never be prime. In fact, one pair of factors of the answer will differ by $4n$ where n is the number originally chosen.

Examples: Let $n = 5$.

$5^4 + 4 = 629 = 37 \times 17$

$37 - 17 = 20 = 4 \times 5$

Let $n = 17$.

$17^4 + 4 = 83{,}525 = 325 \times 257$

$325 - 257 = 68 = 4 \times 17$

5

5 is the 5th Fibonacci number.

5 divides at least one of the elements of every Pythagorean triple.

Examples: Consider these triples

(3,4,5) (12,35,37)

(5,12,13) (9,40,41)

(8,15,17) (28,45,53)

(7,24,25) (33,56,65)

(20,21,29)

6

Every power of 6 ends in 6.

Every prime (except 2 and 3) differs from a multiple of 6 by 1.

6 is the only integer whose divisors (excluding 6) both add and multiply to equal 6; that is,

$$1 + 2 + 3 = 1 \times 2 \times 3 = 6$$

7

Pick any power of 2 and divide it by 7. The remainder is always a power of 2; that is, 1, 2, or 4. In other words,

$$2^n = 1, 2, \text{ or } 4 \pmod 7$$

Note that the sum of the possible remainders is $1 + 2 + 4 = 7$.

8

Select any triangular number, multiply it by 8, and add 1. The result is a square.

Examples: The triangle of 6 = 21

$(21 \times 8) + 1 = 169 = 13^2$

The triangle of 217 = 23,653

$(23{,}653 \times 8) + 1 = 189{,}225 = 435^2$

9

Select any whole number. Scramble its digits in any manner. Subtract the smaller number from the larger one. The difference is always divisible by 9.

Example: Select 17452. Rearrange to 47251.

47251 - 17452 = 29799

29799 ÷ 9 = 3311

10

Despite the fact that 10 is the base of our number system, no power of any number ever ends in 10.

11

11 is the only 2-digit palindrome with a palindromic square, cube, and 4th power.

$11^2 = 121$ $11^3 = 1331$ $11^4 = 14641$

12

The square of 12 is 144. The digits of 144 reversed are 441--another perfect square. Moreover, the square root of 441 is 21, the reverse of 12.

13 and its reverse, 31, also have squares with reversed digits; that is, $13^2 = 169$ and $31^2 = 961$.

13

Consider the set of prime palindromes with a digit

sum of 13 and all digits nonzero. There are exactly 13 such numbers! They are:

373	1411141
12721	3211123
14341	112212211
31513	113111311
32323	121131121
1117111	131111131
1215121	

14

The first 6 digits of the reciprocal of 14 are .071428. Look at these digits in pairs; 07, 14, and 28. Note that

$07 = 14 \times 2^{-1}$ $\quad 14 = 14 \times 2^{0}$ $\quad 28 = 14 \times 2^{1}$

15

15 is the smallest odd composite number that is not a square.

16

16 is the only integer solution of the equation $c = a^b = b^a$ with $a \neq b$ and a and b integers; that is,

$2^4 = 4^2 = 16$

The product of any two consecutive triangular numbers multiplied by 16 and increased by 1 is a square.

Example: The triangles of 13 and 14 are 91 and 105.

$(91 \times 105 \times 16) + 1 = 152881 = 391^2$

17

The difference between any square number and the nearest multiple of 17 is always a power of 2 (or 0).

Examples: Select 13. $13^2 = 169$.

170 (17×10) is nearest multiple of 17 and

$170 - 169 = 1 = 2^0$.

Select 157. $157^2 = 24,649$.

24,650 is the nearest multiple of 17

(17×1450) and $24,650 - 24,649 = 1 = 2^0$.

18

Select any prime greater than 3 and cube it. 18 will divide the result increased by 1 or decreased by 1.

Examples: Select 29.

$29^3 + 1 = 24,390$; $24,390 \div 18 = 1355$

Select 103.

$103^3 - 1 = 1,092,726$; $1,092,726 \div 18 = 60,707$

19

If the numerals of 19 are inverted, the resulting number is 61. Now 19 and 61 are both primes. There exist only 2 other such pairs of numbers with all different digits. They are: 109 and 601; 6091 and 1609.

20

Multiply any prime by 20, add 1, raise to the 4th power, and subtract 1. The result will have at least 80 divisors.

Example: Select 17.

$(17 \times 20) + 1 = 341$

$341^4 - 1 = 13{,}521{,}270{,}960$

$= 2^4 \times 3^2 \times 5 \times 17 \times 19 \times 53 \times 1097$

This number has 480 divisors!

21

Select any multiple of 21. Delete the units digit and multiply by 11. Subtract the units digit that was deleted. The result is divisible by 21.

Example: Select 1743, which is 21×83.

$174 \times 11 = 1914$

$1914 - 3 = 1911$

$1911 \div 21 = 91$

22

Select any number with an even number of digits such that the first and last digits are both odd or both even. Reverse the digits of the number and add the new number to the original number. The sum is divisible by 22.

Example: Select 2768.

$2768 + 8672 = 11{,}440$

$11{,}440 \div 22 = 520$

23

23 is the first prime consisting of two consecutive digits, 2 and 3. Such primes are quite rare. There are two more such 2-digit primes, 67 and 89. No 3-digit primes consisting of 3 consecutive numbers exist. Only one such 4-digit prime exists; it is 4567. The next such prime is the 8-digit number 23,456,789.

23 divides the number 23,571,113. The interesting thing about this number is that its consecutive digits taken singly and in pairs, as appropriate, are the consecutive primes 2 through 13; that is, 2, 3, 5, 7, 11, and 13.

23 also divides 423,571,113,171,923. Except for the first digit, this number is comprised of all the primes from 2 through 23.

24

Select any prime greater than 3. Square it and subtract 1. The result is divisible by 24.

Example: Select 37.

$$37^2 - 1 = 1368; \ 1368 \div 24 = 57$$

25

25 is the smallest square equal to the sum of two other squares; that is,

$$3^2 + 4^2 = 5^2$$

The reverse of 25 is 52 and 52 - 25 = 27, which is 3^3. The next such number to differ from its reverse by a cube is 2031; that is,

$$2031 - 1302 = 729 = 9^3$$

26

26 - 1 = 25, a square; and 26 + 1 = 27, a cube. No other number (less than 10 quadrillion, at least) is squeezed between a square and a cube.

27

27 is the smallest integer that can be expressed as the sum of 3 squares in two different ways.

27 = 25 + 1 + 1

27 = 9 + 9 + 9

28

28 is the smallest integer that can be expressed as the sum of 4 squares in three different ways.

28 = 25 + 1 + 1 + 1

28 = 9 + 9 + 9 + 1

28 = 16 + 4 + 4 + 4

28 is a perfect number; that is, it is equal to the sum of its divisors (excluding 28 itself).

1 + 2 + 4 + 7 + 14 = 28

28 is also a triangular number. As a matter of fact, all perfect numbers are also triangular numbers.

29

Select any whole number. Square it. Find the differences between this square and the two nearest multiples of 29 on either side of it. One of these two differences (if not 0) will divide 1008.

Example: Select 159.

$159^2 = 25,281$

The multiples of 29 on either side are 25,259 and 25,288 with differences from 25,281 of 22 and 7. Either 22 or 7 divides 1008.

It is 7. 1008 ÷ 7 = 144.

30

Every prime number--except for 2, 3, and 5--differs from the nearest multiple of 30 by 1, 7, 11, or 13.

31

The cube root of 31 is 3.1413806. This differs from π = 3.1415927 by less than 0.0067%. The 9th root of 29,809 provides a better approximation of π, namely 3.14159149 but, of course, 29,809 is much larger than 31.

Cube any integer. Find the difference between the cube and the nearest multiple of 31 that differs by an even number. That difference is always a power of 2 (or 0).

Example: Select 176.

$176^3 = 5,451,776$

The nearest multiple of 31 is 5,451,784; the difference is 8, which is 2^3.

32

Multiply 32 by any odd number and square the product. Subtract 1024. The result will always be divisible by 2, 4, 8, 16, 32, 64, 128, 256, 512, 1024, 2048, 4096, and 8192.

Example: Select 7.

$(7 \times 32)^2 = 50,176$

$50,176 - 1024 = 49,152$

$49,152 = 3 \times 2^{14}$

33

Square 33 and reverse its digits. The result is the square of 99.

$33^2 = 1089$ and $9801 = 99^2$

34

No power of any number ever ends in 34.

35

35 is one of only two numbers ending in 5 that is the product of a pair of twin primes; that is, $35 = 5 \times 7$. The other such number is 15.

36

36 divides the product of any two twin primes increased by 1. (The primes must be greater than 3 and 5.) The other factor--other than 36--is a square.

Example: $(41 \times 43) + 1 = 1764$

$1764 \div 36 = 49$

$49 = 7 \times 7$

37

Select any multiple of 37 with 3, 6, 9, 12, etc. digits. The difference between any two cyclic permutations of this number will also be divisible by 37.

Example: Select 123,469, which is 37 × 3337.

346,912 - 234,691 = 112,221

112,221 ÷ 37 = 3033

38

38^2 = 1444. The square of no other number less than 462 ends in three 4s. (462^2 = 213,444.)

39

The repetend of 1/39 is 025641. The first three digits of the repetend added to the last three digits yields 025 + 641 = 666. The repetend of the reciprocal of many prime numbers, when similarly split, yields a sum each digit of which is 9. (For example, 1/137 = .00729927, and 0072 + 9927 = 9999.) However, few repetends of composite numbers (such as 39) combine to yield all 6s.

40

40 is the smallest number ending in 0 that can be multiplied by the square of another number such that the product is a cube; that is,

$$40 \times 5^2 = 1000 = 10^3$$

41

41 is the smallest number with 5 digits in the repetend of its reciprocal; that is, 1/41 = 0.024390243902439.... Two other prime numbers with only 5 digits in the repetend of their reciprocals are 271 and 11,111. (1/271 = .00369 and 1/11,111 = .00009.)

42

Select any prime number greater than or equal to 11 and multiply it by 42. Square the product. The result will always have exactly 81 divisors.

Example: Select 19.

$$(19 \times 42)^2 = 636{,}804$$

Exactly 81 numbers (including 1 and 636,804) divide this result.

The digits of 43 reversed are 34. The 42-digit number

343434343434343434343434343434343434343434

is divisible by 43. In case you're interested, the quotient is 7,986,845,196,147,521,728,917,077,754,287,056,612,638.

44

44 is the only integer whose square is equal to a year in the current 20th century. $44^2 = 1936$.

45

45 is both a triangular number and a hexagonal number. It is the triangle of 9 and the hexagon of 5. Interestingly enough, the product of its triangular root (9) and its hexagonal root (5) is 45.

46

46 divides 9,123,456,789,101,112. Note that, except for the first digit, the number consists of each of the counting numbers, in order, from 1 to 12. When 9,123,456,789,101,112 is divided by 46, the quotient is 1,983,360,117,154,372.

47

There are only two 4-digit numbers that are divisible by 47 whose middle digits are 47. Those two numbers are 3478 and 7473. Interestingly enough, 3478 ÷ 47 = 74, which is the reverse of 47. The other number, 7473, has the reverse of 47 for its two leading digits.

48

Raise any prime (except 2 or 3) to the 4th power and subtract 1. 48 divides the result.

Examples: $7^4 - 1 = 2400$

$2400 \div 48 = 50$

$19^4 - 1 = 130,320$

$130,320 \div 48 = 2715$

49

Add 49 to the smaller of any pair of twin primes (except 3) and square the result. The answer will be divisible by 36 and will have at least 27 divisors.

Example: Select 29.

$$(49 + 29)^2 = 6084; \quad 6084 \div 36 = 169$$

6084 has exactly 27 divisors.

50

50 is the smallest number that can be expressed as the sum of 2 squares in exactly two ways.

$50 = 5^2 + 5^2$ and

$50 = 7^2 + 1^2$

Two other 2-digit numbers can also be expressed as the sum of 2 squares in two ways. They are

$65 = 1^2 + 8^2 = 7^2 + 4^2$ and

$85 = 2^2 + 9^2 = 6^2 + 7^2$

51

Every power of 51 ends in 51 or in 01.

Examples: $51^2 = 2601$

$51^3 = 132,651$

$51^4 = 6,765,201$

$51^5 = 345,025,251$

Select any 6-digit number with its last three digits equal to its first three digits. Multiply that number by 51; then annex a 0 on the right end. The result will be divisible by every prime number from 2 through 17.

Example: Select 589,589.

$51 \times 589{,}589 = 30{,}069{,}039$

$300{,}690{,}390 = 2 \times 3 \times 5 \times 7 \times 11 \times 13$

$\times\ 17 \times (19 \times 31)$

52

The smallest square whose digits add to 52 is 4,999,696, which is 2236^2. Interestingly enough, 2236 is a multiple of 52; that is, $2236 \div 52 = 43$.

Select any 3-digit multiple of 4. Divide it by 999 and keep the first 6 digits of the decimal answer. (Do not round up.) Treat this 6-digit number as a whole number. It is divisible by 52.

Example: Select 156.

$156 \div 999 = .156156...$

$156{,}156 \div 52 = 3003$

53

53 is the smallest power to which 2 can be raised to yield a number beginning with 9; that is,

$$2^{53} = 9{,}007{,}199{,}254{,}740{,}992$$

is the smallest power of 2 beginning with a 9. Powers of 2 beginning with 7 are also a bit difficult to find. The smallest such number is $2^{46} = 70{,}368{,}744{,}177{,}664$.

On the other hand, powers of 2 beginning with 1 are plentiful. In fact, for any given *n*, there is exactly one *n*-digit number that is a power of 2 and begins with 1. For example,

$16 = 2^4$ $\qquad$ $16{,}384 = 2^{14}$

$128 = 2^7$ $\qquad$ $131{,}072 = 2^{17}$

$1024 = 2^{10}$

54

The reciprocal of 54 is 1/54 = 0.0185185185.... The repetend is 185. This number and its cyclical permutations are interesting because

$185 = 37 \times 5$

$851 = 37 \times 23$

$518 = 37 \times 14$

When 54 is divided by this common factor, 37, the quotient is

$$\frac{54}{37} = 1.459459459\ldots$$

This repetend, 459, and its cyclical permutations are

$459 = 27 \times 17$

$594 = 27 \times 22$

$945 = 27 \times 35$

This common factor, 27, is half of 54. Moreover, the product of the two common factors, 37 and 27, is rather interesting because $37 \times 27 = 999$.

55

Select any group of digits and form a large number from these digits as follows.

(a) The left "half" of the numeral formed can be any arrangement of any of the chosen digits.

(b) The right "half" of the numeral must be the digits of the left "half" in reverse order.

(c) Unless the number so formed ends in 5, multiply the result by 10.

This number will be divisible by 55.

Example: Select 2113 and form the number 21133112.

211331120 ÷ 55 = 3,842,384

56

To the nearest hundredth, $\cos 56^{\circ} = .56$.

To be a bit more precise, $\cos 55.96701235^{\circ} = .5596701235$.

57

Raise 8 to any even power. Raise 11 to the same power. The difference between these two results is divisible by 57.

Examples: $11^4 - 8^4 = 10,545$; $10,545 \div 57 = 185$

$11^8 - 8^8 = 197,581,665$

$197,581,665 \div 57 = 3,466,345$

58

Add 58 to the smaller of any pair of twin primes except 3. Square this sum and subtract 9. The result will always be divisible by 36. Upon dividing by 36, the quotient will always be composite with a pair of consecutive integers as one pair of factors.

Example: Select 29 from the twin pair of primes 29 and 31.

$29 + 58 = 87;\ 87^2 - 9 = 7560$

$7560 \div 36 = 210$

$210 = 14 \times 15$

59

59 is the smallest number such that if it is divided by 2, 3, 4, 5, or 6, the remainder will be 1, 2, 3, 4, or 5, respectively.

60

60 is the smallest number with 12 divisors. Those divisors are 1, 2, 3, 4, 5, 6, 10, 12, 15, 20, 30, and 60. There are four other 2-digit numbers with 12 divisors. They are 72, 84, 90, and 96. No 2-digit number has more than 12 divisors.

The repetend of the reciprocal of 61, treated as an integer, is

16,393,442,622,950,819,672,131,147,540,983,606,
557,377,049,180,327,868,852,459

At a glance it might be difficult to decide if this number is prime or not. It most assuredly is not prime. This odd number with 59 digits has an astounding 1,048,576 divisors! Among the more than a million divisors are such numbers as:

9; 99; 999; 9999; 99,999; 999,999; 9,999,999,999;
999,999,999,999; 999,999,999,999,999;
99,999,999,999,999,999,999; and
999,999,999,999,999,999,999,999,999,999

The prime factors of the number are

$$3^3 \times 7 \times 11 \times 13 \times 31 \times 37 \times 41 \times 101 \times 211 \times 241 \times 271$$
$$\times 2161 \times 3541 \times 9091 \times 9901 \times 27{,}961 \times 2{,}906{,}161$$
$$\times 4{,}188{,}901 \times 39{,}526{,}741$$

Moreover, the reciprocal of each of these prime factors has a repetend such that the number of digits in the repetend divides 60. For example, the repetend of the reciprocal of 7 is 142857. This is a 6-digit number and 6 divides 60.

61 is a divisor of 23,571,113,171,923. Note that this number consists of all the primes, in order, from 2 through 23.

$$23{,}571{,}113{,}171{,}923 \div 61 = 386{,}411{,}691{,}343$$

Pick any two consecutive odd numbers, neither of which is divisible by 3. (They must be greater than 5 and 7.) Add their sum to their product and subtract 62. The resulting odd number will always be composite. Moreover, it can always be factored into two numbers that differ by 16. The larger of these two factors will be a multiple of 3; the smaller will be one less than a multiple of 6.

Example: Select 17 and 19.

$$(17 + 19) + (17 \times 19) - 62 = 297$$

$$297 = 27 \times 11 \text{ and } 27 - 11 = 16$$

27 is a multiple of 3, and 11 is 1 less than a multiple of 6.

63

The notation for 63 in bases 2, 4, and 8 is quite interesting.

$$63 = 111111_2 = 333_4 = 77_8$$

64

64 is the only 2-digit number that is a square, a cube, and a sixth power.

$$64 = 8^2 = 4^3 = 2^6$$

Multiply 64 by 4. Then multiply that product by 4, and that product by 4, etc. List each product by shifting the digits 2 places to the right as shown below and add the products. Continue indefinitely. The sum converges to a string of 6s.

```
64
 256
  1024
    4096
     16384
       65536
        262144
         1048576
          4194304
           16777216
-------------------
66666666666665967616
```

65

The reciprocal of 65 is 0.0153846153846.... Considering the repetend as a whole number, it is 153,846. This number factors to $2 \times 3^3 \times 7 \times 11 \times 37$. The prime factors of 65 are 5×13. Thus the prime factors of 65 and the repetend of its reciprocal include each of the first six prime numbers; 2, 3, 5, 7, 11, and 13.

66

Select any 6-digit number such that

(a) the first digit is even.
(b) the first 3 digits are in arithmetic progression (such as 234 or 468, etc.).
(c) The last 3 digits are in reverse order of the first 3 digits.

Any such number is divisible by 66.

Examples: 234,432 ÷ 66 = 3552

258,852 ÷ 66 = 3922

67

Select a 4-digit number whose first 2 digits are equal to twice its last 2 digits. That number is divisible by 67.

Examples: 4824 ÷ 67 = 72

6030 ÷ 67 = 90

7638 ÷ 67 = 114

68

Select any number ending in 8 and raise it to the 5th power. The final 2 digits will be 68.

Examples: 18^5 = 1,889,568 28^5 = 17,210,368

69

69 is the largest number with a factorial of fewer than 100 digits. 69! has 99 digits.

70

70 divides any number of the form
$2(5^{n+2}) + 5(2^{n+5})(-1)^n$.

Examples: If $n = 0$, $2(5^2) + 5(2^5) = 210$; $210 \div 70 = 3$

If $n = 1$, $2(5^3) - 5(2^6) = -70$; $-70 \div 70 = -1$

If $n = 2$, $2(5^4) + 5(2^7) = 1890$; $1890 \div 70 = 27$

If $n = 3$, $2(5^5) - 5(2^8) = 4970$; $4970 \div 70 = 71$

71

The cube of 71 is 357911. Note the progression of digits: 3, 5, 7, 9, 11.

72

Select any pair of twin primes greater than 20. Multiply the two primes, square the product, and subtract 1. The result has at least 72 divisors, one of which is 72.

Example: Select 29 and 31.

$29 \times 31 = 899$; $899^2 = 808{,}201$

$808{,}201 - 1 = 808{,}200$

$808{,}200 = 2^3 \times 3^2 \times 5^2 \times 449$

808,200 has $4 \times 3 \times 3 \times 2$ divisors, one of which is 72.

Multiply 72 by any triangular number and add 9. The result is a square.

Example: Select 28, the triangle of 7.

$(72 \times 28) + 9 = 2025 = 45^2$

73

Select any 8-digit number whose first four digits equal its last four digits. That number is divisible by 73.

Examples: 12341234 ÷ 73 = 169,058

30053005 ÷ 73 = 411,685

The square of 73 is 5329. Break this number into pairs of digits, 53 and 29. The original number 73, along with 53 and 29, are all primes. The next number with similar qualities is 769. The square of this prime is 591361. 769, 59, 13, and 61 are all primes.

74

Select an even number subject to the following conditions:

(a) It may have 3, 6, 9, 12, etc. digits.
(b) Each group of 3 digits repeats that digit 3 times.

74 divides any such number.

Examples: 333,888 ÷ 74 = 4512

222,111,666 ÷ 74 = 3,001,509

75

Select any three whole numbers of any size that are in arithmetic progression. Using all of the digits of these three numbers in any permutation whatever, form a new number. Annex either 00 or 75 at the right end of this number. 75 will divide the result.

Examples: Select 37, 41, and 45.

A permutation of these digits is 7 1 5 4 4 3.
Form the number 71,544,300.
71,544,300 ÷ 75 = 953,924

Select 97, 716, and 1335 (a progression with common difference 619).

From a possible permutation we have 95,737,316,100.
95,737,316,100 ÷ 75 = 1,276,497,548

76

Every power of 76 ends in 76.

Examples: $76^2 = 5776$

$76^3 = 438,976$

$76^4 = 33,362,176$

The following table shows that 76 is just part of a sequence of numbers with the property that every power of the number ends in that number.

EVERY POWER OF	ENDS IN
6	6
76	76
376	376
9,376	9,376
09,376	09,376
109,376	109,376
7,109,376	7,109,376
87,109,376	87,109,376

77

77 is the only composite number less than 187 that ends in 7 and is not divisible by 3.

Select any whole number. Generate two numbers from your selection, one of which is 9 times the number selected, and the other of which is twice the number selected. Square these two numbers and find their difference. The result is always divisible by 77.

Example: Select 17.

$9 \times 17 = 153$ and $2 \times 17 = 34$

$153^2 - 34^2 = 23{,}409 - 1{,}156 = 22{,}253$

$22{,}253 \div 77 = 289$

78

The number of divisors of any power of 78 is equal to a cube.

Examples: 78^2 has 27 divisors.

78^3 has 64 divisors.

78^4 has 125 divisors.

In general, 78^n has $(n + 1)^3$ divisors.

Such a property is certainly not unique to 78. Four other 2-digit numbers also have such a property. They are 30, 42, 66, and 70. Nevertheless, it is a rather surprising property.

79

Select two consecutive odd numbers larger than 79, neither of which is divisible by 3. Multiply the numbers and subtract 79. Divide by 4 and subtract

79 again. Divide by 9 and subtract 79. Finally, subtract 79 one more time. The result will always be composite and can be factored into two factors such that the larger exceeds the smaller by 26.

Example: Select 119 and 121.

$119 \times 121 = 14{,}399$; $14{,}399 - 79 = 14{,}320$

$14{,}320 \div 4 = 3580$; $3580 - 79 = 3501$

$3501 \div 9 = 389$; $389 - 79 = 310$

$310 - 79 = 231$; $231 = 7 \times 33$

7 and 33 differ by 26.

80

80 is the largest 2-digit number with a terminating reciprocal.

$1/80 = 0.0125$

81

$81 = 9 \times 9$. If its digits are reversed, then $18 = 9 + 9$. No other number in base 10 has this property.

81 is the only 2-digit number that is equal to the square of the sum of its digits.

$(8 + 1)^2 = 81$

82

Add 82 to the smaller of a pair of twin primes. Square the result. Subtract 225. This answer is always divisible by 36. Moreover, after dividing by 36, the quotient can be factored into a pair of numbers that differ by 5.

Example: Start with the twin primes 107 and 109.

$107 + 82 = 189$

$189^2 = 35,721$

$35,721 - 225 = 35,496$

$35,496 \div 36 = 986$

$986 = 29 \times 34$

83

The reciprocal of 83 is a repeating decimal whose first several digits are 0.012048192768.... This reciprocal can be generated by beginning with the first three digits, 012, and multiplying by 4. Multiply that product by 4, and by 4 again, etc.

Thus 012 × 4 = 048; 048 × 4 = 192; 192 × 4 = 768; 768 × 4 = 3072; etc.

List each product, as shown in the following sequence, one under the other, but shift each product 3 places to the right of the previous product. Then add.

```
012
   048
      192
         768
         3072
           12288
              49152
              196608
                 786432
0120481927710843373493944 32
```

The sum is approaching the reciprocal of 83 as a limit. The more terms added, the better the approximation.

84

Pick any whole number and follow this symmetric pattern.

DOUBLE, DOUBLE
ADD ORIGINAL
DOUBLE, DOUBLE
ADD ORIGINAL
DOUBLE, DOUBLE

The result is divisible by 84.

Example: Select 13. Here is the sequence of results.

13 26 52 65 130 260 273 546 1092

$1092 \div 84 = 13$

85

Consider the following facts.

$85^2 = 7225$; $85^3 = 614{,}125$; $85^4 = 52{,}200{,}625$;

$85^5 = 4{,}437{,}053{,}125$; $85^6 = 377{,}149{,}515{,}625$

The final 2 digits of 85^2 are $25 = 5^2$.

The final 3 digits of 85^3 are $125 = 5^3$.

The final 3 digits of 85^4 are $625 = 5^4$.

The final 4 digits of 85^5 are $3{,}125 = 5^5$.

The final 5 digits of 85^6 are $15{,}625 = 5^6$.

No other number less than 100 has this property. However, 165, 325, and 645 will extend the pattern even further.

86

The sum of the squares of the digits of 86 is $8^2 + 6^2 = 100$. Except for 68, no other 2- or 3-digit number (if we exclude 0s in such numbers as 608) has the property that the sum of the squares of the digits equals 100. There are, however, some 4-digit numbers such as 5555 and permutations of 1177.

87

Select any 2-digit number. Reverse its digits and add them to the original number. Divide the sum by 0.125. Subtract the sum of the digits of the number originally chosen. The result is divisible by 87.

Example: Select 48. 48 + 84 = 132

132 ÷ 0.125 = 1056

1056 - (4 + 8) = 1044

1044 ÷ 87 = 12

88

The square of 88 is 7744, a number with two pair of equal digits. No other 2-digit number has such a square.

89

89 is a Fibonacci number and its reciprocal, 1/89, can be formed using the Fibonacci sequence. To form this decimal value, list the Fibonacci numbers as

shown below, one under the other, but shift each number one place to the right. Then add.

```
0
 1
  1
   2
    3
     5
      8
      13
       21
        34
         55
          89
          144
-------------
0112359539534
```

This sum is approaching the repetend of 1/89 = 0.0112359551...

Related to this fact are the interesting repetends of 9899 and 998999.

1/9899 = .00010102030508132134 55...

1/998999 = .000001001002003005008013021034055089144 233377610987...

(In 1/998999, note that each group of three digits forms a Fibonacci number!)

90

90 can be expressed as powers of 10 in a rather interesting manner.

$$90 = \frac{10^2}{10^0 + 10^{-1} + 10^{-2} + 10^{-3} + 10^{-4} + \cdots}$$

$$= \frac{100}{1 + 0.1 + 0.01 + 0.001 + 0.0001 + \cdots}$$

91

91 divides any 6-digit number whose first three digits equal its last three digits.

Examples: 371,371 ÷ 91 = 4081

447,447 ÷ 91 = 4917

92

Multiply 92 by 8; then that product by 8; that product by 8, etc. List the products one under the other, shifting the digits two places to the right as shown below. Continue indefinitely. Add. The sum converges to a string of 9s.

```
92
 736
  5888
   47104
    376832
     3014656
      24117248
       192937984
        1543503872
         12348030976
--------------------
9999999999...
```

93

Select any odd multiple of 100 (such as 700 or 2900 or 17789300). Add 93 and square the new number. Subtract 49 and finally divide by 200. The resulting number will be composite. It can be factored into two factors such that the larger one ends in 93. The smaller factor will always be even and the difference between the two factors is divisible by 7.

Example: Select 2900.

$2900 + 93 = 2993$

$2993^2 = 8,958,049$

$8,958,049 - 49 = 8,958,000$

$8,958,000 \div 200 = 44,790$

$44,790 = 1493 \times 30$

$1493 - 30 = 1463$; $1463 \div 7 = 209$

94

Any square number less than 1,000,000 with its first two leading digits 94 will have an even number as its third digit and the remaining digits will equal a square number.

Examples: $97^2 = 9409$; 0 is even and 9 is a square.

$307^2 = 94,249$; 2 is even and 49 is a square.

$971^2 = 942,841$; 2 is even and 841 is a square.

95, of course, is not divisible by 9. Consider any number consisting solely of at least one each of 5s and 9s. No such number less than 5-1/2 billion will be divisible by 9. In fact, the smallest such number is 5,555,555,559.

96

If an integer, *k*, is multiplied by 96, very often (though not always) the product will differ from the next perfect square by a square. This is especially likely to happen if *k* is less than 100.

Example: If $k = 17$, then $17 \times 96 = 1632$; the next square after 1632 is $1681 = 41^2$, and $1681 - 1632 = 49$, also a square.

In the case of an exception, the subsequent perfect square minus $96k$ will frequently be a square.

Example: Let $k = 37$. $96 \times 37 = 3552$. 3600 is the next square, but $3600 - 3552 = 48$ is not a square. However, the next square is $61^2 = 3721$ and $3721 - 3552 = 169 = 13^2$.

97

97 is the only prime number in the 90s. No other decade less than 100 has only one prime in the decade. However, the decades from 110 to 119 and from 120 to 129 each contain only one prime.

98 is the second number of an infinite sequence of numbers none of which is prime. All numbers in the following sequence are composite.

9 98 987 9876 98765 987654 9876543

98765432 987654321 9876543210 98765432109

987654321098 ... 98765432109876543210987654321 0987....

The repetend of the reciprocal of 98 begins with .010204081632.... Note that each pair of digits is twice the pair on its left. The entire repetend can be generated as follows:

```
01
  02
    04
      08
        16
          32
            64
             128
               256
                 512
                  1024
                    2048
                      4096
______________________________
0102040816326530612244896  etc.
```

Similarly, the repetend of 1/998 = 0.001002004008016032064128256.... In this case, each triplet of digits is twice the triplet on its left.

99

Select any multiple of 99. 99 will also divide that number with its digits reversed. Moreover, if the number is broken up into pairs of two digits, then 99 divides any permutation of those 2-digit pairs reassembled into a new number. If all of these pairs are reversed, 99 also divides all reassembled permutations of these reversed digits. (If the number has an odd number of digits such as 12276, treat it as 012276 when breaking it up into pairs.)

Example: 99 × 5229 = 517671. 99 divides its reverse, 176715. The pairs of digits are 51, 76, and 71. Thus, 99 divides all permutations of these pairs reassembled to

517176 715176 717651 765171 767151

The reverses of the pairs are 17, 67, and 15. 99 also divides

171567 151767 156717 671517 671715

99 divides the difference between the square of any integer and the square of its reverse.

Example: Select 753.

$753^2 - 357^2 = 439{,}560$

$439{,}560 \div 99 = 4440$

Select any whole number with an even number of digits. Reverse the order of the digits and add to the original number. Reverse the digits of the result and subtract from the result. The answer is divisible by 99.

Example: Select 1983.

$1983 + 3891 = 5874$

$5874 - 4785 = 1089$

$1089 \div 99 = 11$

100

100 is, of course, the smallest number with three digits. The value of 100^n (for n a positive integer) is always the smallest number with $(n + 1)$ digits. Moreover, the square root of 100 is the smallest number with two digits.

101

If two integers add to 101, then the difference of their squares is equal to their difference "repeated."

Examples: $79 + 22 = 101$

$79^2 - 22^2 = 5757$

$79 - 22 = 57$

$45 + 56 = 101$

$56^2 - 45^2 = 1111$

$56 - 45 = 11$

102

The powers of 102 up to and including 102^5 display an especially intriguing pattern.

102^2 = 10404, which groups into pairs of 1, 04, and 04. A square consists of 1 square, 4 edges, and 4 vertices.

102^3 = 1061208, which groups into pairs of 1, 06, 12, and 08. A cube consists of 1 cube, 6 square faces, 12 edges, and 8 vertices.

102^4 = 108243216, which groups into pairs of 1, 08, 24, 32, and 16. A 4-dimensional tesseract consists of 1 tesseract, 8 cubes, 24 squares, 32 edges, and 16 vertices.

102^5 = 11040808032, which groups into pairs of 1, 10, 40, 80, 80, and 32. A 5-dimensional hypercube consists of one 5-dimensional hypercube, 10 tesseracts, 40 cubes, 80 squares, 80 edges, and 32 vertices.

For powers of 102 greater than 102^5, "carrying" destroys the pattern. Powers of 1002 will carry the pattern further.

103

Every power of 103 differs from a multiple of 13 by 1 and from a multiple of 17 by 1.

Examples: 103^2 = 10,609

816 × 13 = 10,608 and 624 × 17 = 10,608

103^3 = 1,092,727

13 × 84,056 = 1,092,728 and 17 × 64,278 = 1,092,726

104

Cube any even number. Find the difference between this cube and the two multiples of 104 on either side of it. One of these differences will equal either 2^3 or 2^6 (or 0).

Example: $10^3 = 1000$. The multiples of 104 on either side of 1000 are 936 and 1040. $1000 - 936 = 64 = 2^6$.

105

If we define a triplet of primes to be three consecutive prime numbers with a common difference of 2, then 105 is the only number that is the product of a triplet of primes.

$105 = 3 \times 5 \times 7$

106

106 is not, of course, a palindrome. However, when it is multipled by a palindrome, the product is frequently a palindrome.

$106 \times 2 = 212$; $106 \times 4 = 424$; $106 \times 6 = 636$; $106 \times 8 = 848$; $106 \times 22 = 2332$; $106 \times 44 = 4664$; $106 \times 66 = 6996$; $106 \times 202 = 21412$; $106 \times 222 = 23532$; $106 \times 242 = 25652$

It appears that if the palindrome consists of 2s, 4s, and 0s, nearly any such multiple of 106 is a palindrome. For example, $106 \times 24042 = 2548452$ and $106 \times 4240424 = 449484944$.

107

Select any square greater than 107 and add the fraction 2/9. Subtract 107 and convert this mixed number to an improper fraction. The numerator will be a composite number with a pair of factors differing by 62.

Example: Select $13^2 = 169$.

$$169\frac{2}{9} - 107 = \frac{560}{9} = \frac{70 \times 8}{9}$$

$$70 - 8 = 62$$

108

$108 = 2^2 \times 3^3$ and thus is of the form $p^p q^q r^r s^s \ldots$ etc., where p, q, r, and s are consecutive primes beginning with 2. There are, of course, an infinite number of such numbers, but nevertheless they are far between. The next such number is

$$2^2 \times 3^3 \times 5^5 = 337{,}500$$

Following that comes

$$2^2 \times 3^3 \times 5^5 \times 7^7 = 277{,}945{,}762{,}500$$

109

109 or any power of 109 differs by only 1 from the nearest multiple of <u>each</u> of the following numbers.

2 3 4 5 6 9 11 12 18 22 27 36 54 55 108 110

Example: $109^3 = 1{,}295{,}029$; when reduced by 1, we get 1,295,028, which is divisible by 2, 3, 4, 6, 9, 12, 18, 27, 36, 54, and 108.

When 109^3 is increased by 1, we get 1,295,030, which is divisible by the rest of the numbers in the list; that is, by 5, 11, 22, 55, and 110.

110

Select any number with an even number of digits. Reverse the digits and add the two numbers. Call the sum S. From S, subtract any palindrome with an even number of digits such that the last digit of the palindrome is the same as the last digit of S. This difference is divisible by 110.

Example: Select 296,883.

296,883 + 388,692 = 685,575
Pick a palindrome with an even number of digits ending in 5; 5225.
685,575 - 5225 = 680,350
680,350 ÷ 110 = 6185

111

Select any square number, multiply it by 1000, and divide by 111. Keep only the greatest integer in the answer, calling it n. Calculate

$$n - \left[\frac{n}{1000}\right] \quad \text{(where } \left[\frac{n}{1000}\right] \text{ is the greatest integer in } \frac{n}{1000}\text{)}$$

The result will be a square.

Example: Select $39^2 = 1521$.

1,521,000 ÷ 111 = 13702.7027

$n = 13,702$

$$13,702 - \left[\frac{13,702}{1000}\right] = 13,702 - 13$$

$$= 13,689 = 117^2$$

112

Select any odd number and raise it to the 7th power. Then subtract the original number. Divide the result by 1.50. The quotient will be divisible by 112.

Example: Select 17.

$17^7 = 410,338,673$

$410,338,673 - 17 = 410,338,656$

$410,338,656 \div 1.5 = 273,559,104$

$273,559,104 \div 112 = 2,442,492$

113

The cube of 113 is quite interesting. $113^3 = 1,442,897$.

The first 3 digits, 144, $= 12^2$.

The next 3 digits, 289, $= 17^2$.

The final 7 does not appear to make a useful contribution to a pattern.

114

Select any prime number greater than 3 and square it. Subtract 1. Call the result n. If $\left(1/2\right)n$ is subtracted from $10n$, the result is divisible by 114.

Example: Select 13.

$13^2 = 169$

$169 - 1 = 168 = n$

$1680 - 84 = 1596$

$1596 \div 114 = 14$

115

The sum of the squares of the digits of 115 is a cube; that is, $1^2 + 1^2 + 5^2 = 27 = 3^3$. What other 3-digit numbers have this property? (Of course, 151 and 511 have the property.) In addition, there are

202, where $2^2 + 0^2 + 2^2 = 8 = 2^3$

220, where $2^2 + 2^2 + 0^2 = 8 = 2^3$

333, where $3^2 + 3^2 + 3^2 = 27 = 3^3$

568, where $5^2 + 6^2 + 8^2 = 125 = 5^3$

Other combinations of this last number--586, 658, 685, 856, and 865--also have this property. There are no others.

116

116 is the sum of three consecutive even squares; that is,

$$4^2 + 6^2 + 8^2 = 16 + 36 + 64 = 116$$

Of course, there are many numbers that are the sum of three consecutive even squares or three consecutive odd squares. However, if any such number is first reduced by $2^3 = 8$ and that difference multiplied by $3^3 = 27$, the product is a square.

Example: $(116 - 8) \times 27 = 2916 = 54^2$

Consider $17^2 + 19^2 + 21^2 = 1091$.

$(1091 - 8) \times 27 = 29{,}241 = 171^2$

117

List 117 and all its cyclical permutations. Consider also the repetend of the reciprocal of 117, which is .008547. Treat the repetend as if it were a 6-digit whole number, 008547, and consider all of its cyclical permutations. Next, reverse the digits of the repetend and list all of the cyclical permutations of that number. The numbers now listed are:

117	171	711			
8547	547008	470085	700854	85470	854700
7458	800745	580074	458007	74580	745800

The first nine primes are 2, 3, 5, 7, 11, 13, 17, 19, and 23. Each of these primes divides at least one of the listed numbers.

118

Select three numbers, the first a square, the second equal to the first number minus 1, and the third equal to the second number minus 3. Multiply these three numbers and reduce the product by 1 2/3%. The result is divisible by 118.

Example: Select $41^2 = 1681$.

$1681 \times 1680 \times 1677 = 4{,}735{,}982{,}160$

1 2/3% of this number is 78,933,036

$4{,}735{,}982{,}160 - 78{,}933{,}036 = 4{,}657{,}049{,}124$

$4{,}657{,}049{,}124 \div 118 = 39{,}466{,}518$

119

Select any square number and multiply it by 7. The result differs from the nearest multiple of 119 by a composite number, one of whose factors is 7. The other factor is a power of 2 (or 0).

Example: Select $651^2 = 423{,}801$.

$423{,}801 \times 7 = 2{,}966{,}607$

The nearest multiple of 119 is
$119 \times 24{,}929 = 2{,}966{,}551$.
$2{,}966{,}607 - 2{,}966{,}551 = 56 = 7 \times 8 = 7 \times 2^3$.

120

Select any prime greater than 5 and raise it to the 4th power. Subtract 1. 120 divides the result.

Example: Select 17.

$17^4 = 83{,}521$; $83{,}521 - 1 = 83{,}520$

$83{,}520 \div 120 = 696$

121

121 is a palindrome. So is its square, 14641, and so is its square root, 11.

The reciprocal of 121 is 0.0082644628099173353719...
The second through the tenth digits are 0826446280, a palindrome; the 12th through the 22nd digits are 91733533719, also a palindrome.

The reciprocal of 11, the square root of 121, is .09090909... which can be broken into two palindromes, 090 and 909.

Moreover, if 121 is multiplied by the "near palindrome" 918273645546372819l, the product is the palindrome 1,111,111,111,111,111,111,111.

When 121 is expressed in several other bases, the digits form a palindrome in those bases; that is,

$$121_{10} = 171_8 = 232_7 = 11111_3$$

While the base 9 and base 5 representations of 121_{10} are not palindromes, moved together as a 6-digit number they do form a palindrome; that is,

$121_{10} = 144_9 = 441_5$ and 144441 is a palindrome.

122

122 is the second in an infinite set of numbers with a very interesting property. The set is 12, 122, 1222, 12222, 122222, etc.

We first note the squares of these numbers along with the squares of their reverses.

$12^2 = 144$ $\quad 122^2 = 14884$ $\quad 1222^2 = 1493284$

$21^2 = 441$ $\quad 221^2 = 48841$ $\quad 2221^2 = 4932841$

$12222^2 = 149377284$ $\quad 122222^2 = 14938217284$

$22221^2 = 493772841$ $\quad 222221^2 = 49382172841$

Note that the squares of 12, 122, 1222, etc. all begin with 14 and end in 4. Now look at the numbers inserted between the 14 and the 4. They are (starting with the squares of 122 and 221):

$88 = 88 \times 1$; $9328 = 88 \times 106$; $937728 = 88 \times 10656$;

$93821728 = 88 \times 1066156$

Moreover, $21^2 - 12^2 = 297 = 9 \times 33$

$221^2 - 122^2 = 33957 = 99 \times 343$

$2221^2 - 1222^2 = 3439557 = 999 \times 3443$

$22221^2 - 12222^2 = 344395557 = 9999 \times 34443$

123

Form any 6-, 7-, 8-, 9-, or 10-digit number according to the following restrictions.

6-digit $a9999A$; 7-digit $ab999AB$; 8-digit $abc99ABC$;

9-digit $abcd9ABCD$; 10-digit $abcdeABCDE$

In each case, the digits a + A must equal 9; likewise, $b + B = 9$; $c + C = 9$; and $d + D = 9$. Any number so formed is divisible by 123.

Examples: 299,997 ÷ 123 = 2439

2,399,976 ÷ 123 = 19,512

15,799,842 ÷ 123 = 128,454

211,397,886 ÷ 123 = 1,718,682

6,494,835,051 ÷ 123 = 52,803,537

124

$124 = 2^2 + 2^3 + 2^4 + 2^5 + 2^6 = 2^7 - 2^2$.
Of course, there exist an infinity of numbers of the form $a^2 + a^3 + a^4 + a^5 + a^6$ and an infinity of numbers of the form $a^7 - a^2$, but 124 is the only number that can be expressed both ways.

125

Select any odd square and subtract 1. Divide the result by 125 (= 5^3). Ignore the decimal point and treat the entire quotient as if it were a whole number. This number will always be divisible by 64 (= 4^3).

Example: Select 37.

$37^2 - 1 = 1368$

1368 ÷ 125 = 10.944

10,944 ÷ 64 = 171

This first property of 126 is a bit complicated, but the results are surprising. Select any proper or improper fraction of the form $k/3$ and multiply it by 126. (k is a positive whole number.) Either add to this result or subtract from it (you choose) any one of the following numbers.

1 2 3 4 5 8 9 10 15 16 17 20

Cube the result. Next find the two multiples of 126 on either side of this cube and find the difference between each of these multiples and the cube. At least one, if not both, of these differences will also be a cube.

Example: First select $\frac{13}{3}$.

$\frac{13}{3} \times 126 = 546$

Choose 5 from the list above and subtract; 546 - 5 = 541.

$541^3 = 158,340,421$

The "neighboring" multiples of 126 are 158,340,420 and 158,340,546. The two differences are 1 and 125, which are both cubes.

Select any multiple of 126. Mix its digits in any manner, but leave an even number in the units place. Multiply by 7. The procedure may be repeated again and again. Each time after the permuted digits have been multiplied by 7, the result will be divisible by 126.

Example: Select 1638, which is 126 × 13.

Number	Permutation	× 7	÷ 126
1638	3816	26712	212
26712	12276	85932	682
85932	92358	646506	5131

etc.

127

Since $127 = 2^7 - 1$, it might be a good time to look at interesting facts about all numbers of the form $2^n - 1$.

(a) If n is composite, then $2^n - 1$ is composite.

Example: Since 15 is composite, $2^{15} - 1 = 32{,}767$ is composite. In fact, $32{,}767 = 7 \times 31 \times 151$.

(b) If n is composite and $n = rs$ where r and s are integers, then both $2^r - 1$ and $2^s - 1$ divide $2^n - 1$.

Example: Since $35 = 5 \times 7$, then $2^5 - 1$ and $2^7 - 1$ both divide $2^{35} - 1$.

(c) If a divides $2^n - 1$, then a divides $2^{kn} - 1$ where k is an integer.

Example: Since 7 divides $2^6 - 1 = 63$, then 7 also divides $2^{12} - 1$, $2^{18} - 1$, $2^{24} - 1$, etc.

(d) For every prime p greater than 2 there exists an n such that p divides $2^n - 1$. In fact, $n = p - 1$.

Example: 11 divides $2^{10} - 1 = 1023 = 11 \times 93$.

(e) If n is prime, $2^n - 1$ might be prime or composite.

Example: $2^7 - 1 = 127$ is prime, but $2^{11} - 1 = 2047 = 23 \times 89$.
$2^{29} - 1 = 536{,}870{,}911 = 233 \times 1103 \times 2089$.

In the examples involving $2^{11} - 1$ and $2^{29} - 1$, note that the exponent n divides each factor reduced by 1; that is, 11 divides (23 - 1) and (89 - 1), and 29 divides (233 - 1), (1103 - 1), and (2089 - 1).

128

128 has no odd divisors. Only two other 3-digit numbers have no odd divisors; they are 256 and 512.

Because $128 = 2^7$, it might be appropriate at this time to look at properties of other numbers of the form 2^n.

(a) For any positive integer n, there exists exactly one n-digit number of the form 2^k with first digit equal to 1.

No. of Digits	1	2	3	4	5	6	7	8
Power of 2 with 1st Digit 1	1	16	128	1024	16384	131072	1048576	16777216

(b) For any positive integer n, there are at least 3, but no more than 4, n-digit powers of 2.

Examples: The 3-digit powers of 2 are 128, 256, and 512. The 4-digit powers of 2 are 1024, 2048, 4096, and 8192.

(c) A relatively small proportion of powers of 2 begin with 9. The smallest is $2^{53} = 9{,}007{,}199{,}254{,}740{,}992$. Once a power of 2 does begin with a 9, then others "soon" follow.

All powers of 2 beginning with a 9 that are less than 10^{100} follow:

2^{53} 2^{63} 2^{73} 2^{83} 2^{93}

2^{156} 2^{166} 2^{176} 2^{186}

2^{249} 2^{259} 2^{269} 2^{279} 2^{289}

Of the 333 powers of 2 less than 10^{100}, only 14 begin with a 9. On the other hand, 100 of these numbers begin with a 1.

129

$129 = 2^7 + 1$. Just as numbers of the form $2^k - 1$ are of interest (see 127), so also are numbers of the form $2^k + 1$.

Whereas $(2^k - 1)$ is prime <u>only</u> if k is prime, $(2^k + 1)$ is prime <u>only</u> if k is a power of 2.

Example: $2^2 + 1 = 5$; $2^4 + 1 = 17$; $2^8 + 1 = 257$; $2^{16} + 1 = 65{,}537$. Each is a prime.

However, $2^{32} + 1 = 4{,}294{,}967{,}297 = 641 \times 6{,}700{,}417$.

Thus, although k must be a power of 2 for $2^k + 1$ to be prime, $2^k + 1$ is not <u>necessarily</u> prime.

130

We note that $130 = 65 \times 2$ and that $310 = 62 \times 5$. Are there any other 3-digit numbers with the characteristics that if the first two digits are interchanged, then the units digits of the factors can be interchanged?

Yes. $324 = 36 \times 9$ and $234 = 39 \times 6$. There are no others.

131

131 is the first of a string of numbers with squares with an interesting pattern.

$131^2 = 17161$

$1331^2 = 1771561$

$13331^2 = 177715561$

$133331^2 = 17777155561$

$1333331^2 = 1777771555561$ etc.

132

Select any palindrome with an even number of digits. Call it *p*. Form another number *q* from *p* by selecting any cyclical permutation of *p*. Find the positive difference between *p* and *q* and increase the result by 1/3. This number is divisible by 132.

Example: Select *p* to equal 571175 and let *q* = 117557.

$p - q = 453,618$.

Increased by 1/3 we get 604,824.

$604,824 \div 132 = 4582$

133

133 is the sum of the cubes of two different primes; that is, $133 = 2^3 + 5^3$. Only five other numbers less than 1000 have such a property. They are:

$35 = 2^3 + 3^3$; $351 = 2^3 + 7^3$; $152 = 3^3 + 5^3$;

$370 = 3^3 + 7^3$; and $468 = 5^3 + 7^3$

133 can also be expressed as $133 = 6^3 - 4^3 - 3^3 + 2^3$ $= 2^3 + 5^3$. This means that

$$6^3 - 4^3 - 3^3 + 2^3 = 2^3 + 5^3$$

This equation, while not very interesting, does lead to the surprising fact that

$$3^3 + 4^3 + 5^3 = 6^3$$

134

Consider the four consecutive numbers 10, 11, 12 and 13. It happens that $134 = 12^2 - 10 = 11^2 + 13$. In general, if four consecutive numbers are selected, the square of the third minus the first equals the square of the second plus the fourth. Now, if 2 is subtracted from any such number and the result is divided by 2, the answer will be a triangular number. In this case, $(134 - 2) \div 2 = 66$, which is the triangle of 11.

Example: Select the 4 consecutive numbers 47, 48, 49, and 50.

$49^2 - 47 = 48^2 + 50 = 2354$

$2354 - 2 = 2352$; $2352 \div 2 = 1176$,

the triangle of 48.

135

Select any whole number greater than 100. Leaving the units digit in place, mix the remaining digits in any manner. Subtract the smaller of the two numbers from the larger and square the difference. The result is divisible by 135. Moreover, the quotient, when divided by 60, is a perfect square.

Example: Select 64,948. Rearrange to 94,468.

$94{,}468 - 64{,}948 = 29{,}520$

$29{,}520^2 = 871{,}430{,}400$

871,430,400 ÷ 135 = 6,455,040

Moreover, 6,455,040 ÷ 60 = 328^2.

136

Select any whole number, multiply by 10, square, and divide by 12.5. Find the difference between this number and the nearest multiple of 136. The difference is a power of 2 (or 0).

Example: Select 57.

570^2 = 324,900

324,900 ÷ 12.5 = 25,992

The nearest multiple of 136 is 136 × 191 = 25,976. 25,992 - 25,976 = 16 = 2^4

137

Select any 8-digit number with its first four digits the same as its last four digits. Any such number is divisible by 137.

Examples: 47894789 ÷ 137 = 349597

24682468 ÷ 137 = 180164

138

Each of the digits of 138, namely 1, 3, and 8, is a Fibonacci number. The first two digits are 13, another Fibonacci number. Moreover, the sum of the number formed by the first two digits added to the final digit is a Fibonacci number; that is, 13 + 8 = 21. And, of course, 13 - 8 = 5 is another Fibonacci number. The difference between 138 and 144, the next Fibonacci number, is 6, the triangle of 3, another

Fibonacci number. The difference between 138 and 89, the previous Fibonacci number, is 49, the square of 7, which is the sum of the first four Fibonacci numbers; that is, 1 + 1 + 2 + 3 = 7.

139

Even though 139 is prime, any even power of 139 minus 1 is divisible by all of the following numbers!

2 3 4 5 6 7 8 10 12 14 15 20 21 23 24

28 30 35 40 42 46 56 60 69 70 84 92 105

115 120 138 140 161 168 184 210 230 276

280 322 345 420 460 483 552 644 690 805

840 920 966 1288 1380 1610 1932 2415 2760

3220 3864 4830 6440 9660 and 19,320

140

Select any number n from 1 to 41 except 15, 21, and 30. Find the difference between 140 and the multiple of n nearest to 140. Also, find the difference between the square of 140 and the multiple of n nearest to the square of 140. At least one of these two results will be a power of 2.

Example: Let $n = 29$.

The multiple of 29 nearest to the square of 140 or 19,600 is 29 × 676 = 19,604.

19,604 - 19,600 = 4 = 2^2.

141

The consecutive integers 2, 3, 4, and 5 are each used once on the left side of the equation

$4^2 + 5^3 = 141$

142

Select any whole number and decrease it by 0.6%. Ignore the decimal point in the answer, treating it as a whole number. 142 divides this number.

Example: Select 17.

$0.6\% \times 17 = .102$

$17 - .102 = 16.898$

$16898 \div 142 = 119$

143

Select any two twin primes greater than 5 and 7. Multiply them and subtract 143. The result is divisible by 36. Moreover, after dividing by 36, if the quotient is increased by 4, the answer is a square.

Example: Select 41 and 43.

$41 \times 43 = 1763$

$1763 - 143 = 1620$

$1620 \div 36 = 45$

$45 + 4 = 49 = 7^2$

144

Consider the set of all right triangles with integral sides. No such right triangle has a hypotenuse of 144. However, 144 is a leg of 17 different right triangles. It is the smallest number that is the leg of 17 right triangles. Those triangles are:

(144,17,145)	(144,42,150)	(144,60,156)
(144,108,180)	(144,130,194)	(144,165,219)
(144,192,240)	(144,270,306)	(144,308,340)
(144,420,444)	(144,567,585)	(144,640,656)
(144,858,870)	(144,1292,1300)	(144,1725,1731)
(144,2590,2594)	(144,5183,5185)	

Even though 144 is the smallest number that is a leg of 17 triangles, there is a smaller number that is a leg of even more triangles; that is, 120 is a leg of 23 right triangles. Moreover, 120 can be the hypotenuse of a right triangle.

145

Subtract 1 from 145; the result is a square (of 12). Double 145 and subtract 1; the result is also a square, $289 = 17^2$. This fact may not seem very interesting unless we realize that there are only 7 numbers less than 220 billion with the property that the number minus 1 is a square and double the number minus 1 is also a square. Here they are.

n	n − 1	2n	2n − 1
5	2^2	10	3^2
145	12^2	290	17^2
4,901	70^2	9,802	99^2
166,465	408^2	332,930	577^2
5,654,885	$2,378^2$	11,309,770	$3,363^2$
192,099,601	$13,860^2$	384,199,202	$19,601^2$
6,525,731,525	$80,782^2$	13,051,463,050	$114,243^2$

It appears that the ratio of two consecutive such numbers is approaching a limit of 33.97056276... =

$$(1 + \ 2)^4.$$

If this is correct, the next value of n in the table is 221,682,772,225.

146

Select any even 2-digit number. Form an 8-digit number by repeating the selected number 4 times. 146 divides the 8-digit number.

Examples: Select 38.

38383838 ÷ 146 = 262,903

Select 40.

40404040 ÷ 146 = 276,740

147

Select any rational number and subtract the square of the number from the square of its double. Call the result n. Next, find the value of each of the three following expressions.

$$\frac{n}{147} \qquad \frac{147}{n} \qquad 147n$$

Each of these three numbers is a square of a fraction (or whole number).

Example: Select the fraction $\frac{4}{7}$.

$$\left[\frac{8}{7}\right]^2 - \left[\frac{4}{7}\right]^2 = \frac{64 - 16}{49} = \frac{48}{49} = n$$

$$\frac{n}{147} = \frac{48}{7203} = \frac{16}{2401} = \left[\frac{4}{49}\right]^2$$

$$\frac{147}{n} = 150.0625 = 12.25^2$$

$$147n = 147 \times \frac{48}{49} = 144 = 12^2$$

148

Select a whole number that meets the following conditions.

(a) The number of digits in the number is a multiple of 3.
(b) The three numbers in a given triplet of digits are all equal.
(c) The final triplet must be 000, 444, or 888.

Any number that meets all of these conditions is divisible by 148.

Examples: 777,111,444 ÷ 148 = 5,250,753

555,000,444 ÷ 148 = 3,750,003

The six permutations of the digits of 149 are

149 194 419 491 914 941

Consider the 27-digit number whose final 18 digits are, in triplets, the 6 permutations of 149, in order.

100,000,000,149,194,419,491,914,941

This number is divisible by 149; that is,
100,000,000,149,194,419,491,914,941 ÷ 149
= 671,140,940,598,620,265,046,409

150

150 multiplied by its double, 300, is exactly 100 times as great as the sum of the same two numbers; that is,

$$150 \times 300 = 45{,}000 = (150 + 300) \times 100 = 45{,}000$$

A related fact of somewhat greater interest involves the fraction that equals 1% of 150, namely 1.5. This number plus its double equals the product of the same two numbers.

$$1.5 + 3 = 1.5 \times 3 = 4.5$$

151

An interesting fact about 151 is that it is the smallest number for which nothing interesting is given in this book!

****

If we accept the hypothesis that every number is special, then this is no place to stop. It seems clear, though, that there is no point at which it would ever be appropriate to stop. Listed below are several more interesting numbers. Our chore now is to fill in the gaps, at least.

169

Look at the sum of the factorials of the digits of 169.

$$1! + 6! + 9! = 1 + 720 + 362{,}880 = 363{,}601$$

The sum of the factorials of the digits of this number is

$3! + 6! + 3! + 6! + 0! + 1! = 6 + 720 + 6 + 720 + 1 + 1 = 1454$

One more time.

$1! + 4! + 5! + 4! = 1 + 24 + 120 + 24 = 169$,

the number we started with.

209

Any even power of 209 differs by 1 or 0 from the nearest multiple of <u>each</u> of 103 different numbers! These numbers are:

2 3 4 5 6 7 8 10 11 12 13
14 15 16 19 20 21 24 26 28 30
32 35 39 40 42 48 52 56 60 65
70 78 80 84 91 96 104 105 112
120 121 130 140 156 160 168 182 195
208 209 210 224 240 260 273 280 312
336 361 364 390 416 420 455 480 520
546 560 624 672 728 780 840 910
1040 1092 1120 1248 1365 1456 1560
1680 1820 2080 2184 2299 2730 2912
3120 3360 3640 3971 4368 5460 6240
7280 8736 10920 14560 21840 43680 43681

Example: An even power of 209 is $209^4 = 1{,}908{,}029{,}761$. Any number in the list above will either divide this number or this number decreased by 1.

$1{,}908{,}029{,}760 \div 455 = 4{,}193{,}472$

210

Slightly more than 77% of all whole numbers have a factor in common with 210.

216

216, which is the smallest even 3-digit cube, is also the smallest number with 16 divisors. We note further that 16 is both a square and a 4th power.

225

225 is a "popular" polygonal number.

225 is the square of 15.
225 is the octagon of 9.
225 is the 24-gon of 5.
225 is the 76-gon of 3.
225 is the 225-gon of 2.

325

325 is the smallest number that can be expressed as the sum of two squares in three ways.

$$325 = 1^2 + 18^2 = 6^2 + 17^2 = 10^2 + 15^2$$

There are six other 3-digit numbers that can be expressed as the sum of two squares in three ways. They are 425, 650, 725, 845, 850, and 925.

Extending the idea, 1105 is the smallest number that can be expressed as the sum of two squares in four ways.

$$1105 = 4^2 + 33^2 = 9^2 + 32^2 = 12^2 + 31^2 = 23^2 + 24^2$$

Other numbers less than 5000 that can be expressed as the sum of two squares in four ways are:

1625	1885	2125	2210	2405	2465	3245
3250	3445	3485	3625	3770	3965	4225
4250	4420	4505	4625	4810	4930	

32,500 can be expressed as the sum of two squares in five ways. (It might not be the smallest such number.)

$$32{,}500 = 10^2 + 180^2 = 54^2 + 172^2 = 60^2 + 170^2$$
$$= 100^2 + 150^2 = 116^2 + 138^2$$

110,500 can be expressed as the sum of two squares in eight ways! (It might not be the smallest such number.)

$$110{,}500 = 40^2 + 330^2 = 54^2 + 328^2 = 90^2 + 320^2$$
$$= 120^2 + 310^2 = 166^2 + 288^2 = 176^2 + 282^2$$
$$= 202^2 + 264^2 = 230^2 + 240^2$$

371

371 is equal to the sum of the cubes of its digits.

$$3^3 + 7^3 + 1^3 = 27 + 343 + 1 = 371$$

There exist three more 3-digit numbers with the same property. They are 153, 370, and 407.

There is no 3-digit number that is equal to the sum of each of its digits raised to the *n*th power for *n* greater than 3 or *n* less than 3.

541

541 is the 100th prime number. The 1,000th prime is 7919 and the 10,000th prime is 104,729.

561

561 is a rather important configurate number.

561 is the triangle of 33.
561 is the hexagon of 17.
561 is the 12-gon of 11.
561 is the 39-gon of 6.
561 is the 188-gon of 3.
561 is the 561-gon of 2.

840

840 can be the length of one of the legs in 67 different right triangles (with integral sides). No number smaller than 840 appears in nearly as many Pythagorean triples. The previous "record holder" is 420 (which incidentally is 1/2 of 840), which is one of the legs in only 40 Pythagorean triples.

2520

2520 is the smallest number divisible by every number from 1 to 10. The smallest number divisible by every number from 1 to 28 is 80,313,433,200.

45,360

The smallest number with exactly 100 divisors is 45,360.

$$45{,}360 = 2^4 \times 3^4 \times 5 \times 7$$

1,152,921,504,606,646,976

Except for 1, the number 1,152,921,504,606,646,976 (1 quintillion, 152 quadrillion, 921 trillion, 504 billion, 606 million, 646 thousand, 976) is the smallest number that is a perfect square, cube, fourth, fifth, and sixth power. It is also a 10th, 12th, 15th, 20th, 30th, and 60th power.